Pocket Revision: Basic Trigonometry

Alexander Forrest

ISBN 978-0-9576916-2-9

Published in the United Kingdom by:

Beagle Bytes

For

P

CONTENTS

CONTENTS v

Definitions.. 6

Using Tan to find angle 8

Using Tan to find side 10

Using sin to find side.. 12

Using Sin to find angle 14

Using Cos to find side... 16

Using Cos to find angle 18

Which ratio should be used ? 20

Finding the area of a triangle.................................. 24

The Sine Rule 30

The CoSine Rule 35

Which rule to use.. 43

Books by the Author 46

DEFINITIONS

Trigonometry is the study of triangles.

The sides of any right angled triangle can be labelled Hypotenuse, Adjacent and Opposite.

The hypotenuse is always opposite the right angle. It is the longest side of the triangle.

The adjacent is the side that forms part of the required angle.

The opposite is the side directly across from the required angle.

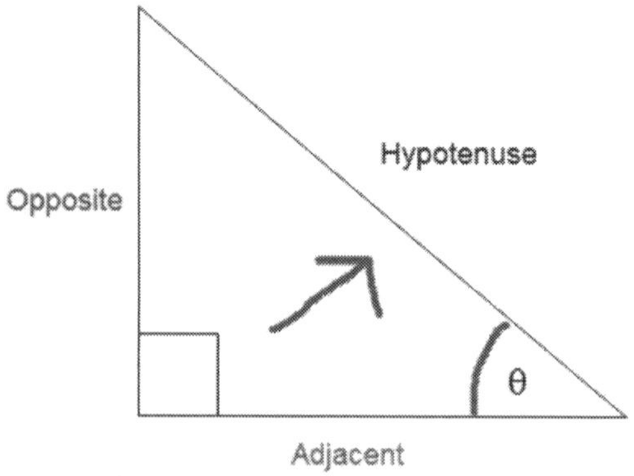

The ratio of these sides is given special names:

sine, cosine and tangent.

These are shortened to sin, cos and tan

$$\text{Sin}\,\theta = \frac{Opposite}{Hypotenuse} \quad \text{remembered as} \quad S\frac{O}{H} \quad or \quad \text{SOH}$$

$$\text{Cos}\,\theta = \frac{Adjacent}{Hypotenuse} \quad \text{remembered as} \quad \text{C}\frac{A}{H} \quad or \quad \text{CAH}$$

$$\text{Tan}\,\theta = \frac{Opposite}{Adjacent} \quad \text{remembered as} \quad \text{T}\frac{O}{A} \quad or \quad \text{TOA}$$

To find an angle or side, follow this recipe:-

- Find and sketch the triangle
- Mark the right angle
- Identify and mark angle to be used / found
- Label Opposite, Hypotenuse and Adjacent
- Write out ratio
- Write down solution

Always draw a sketch

USING TAN TO FIND ANGLE

Example 1

An aircraft is 73m above a building and 200m from its touchdown point.

Calculate angle θ°, the glide path of the aircraft.

Give your answer correct to one decimal place.

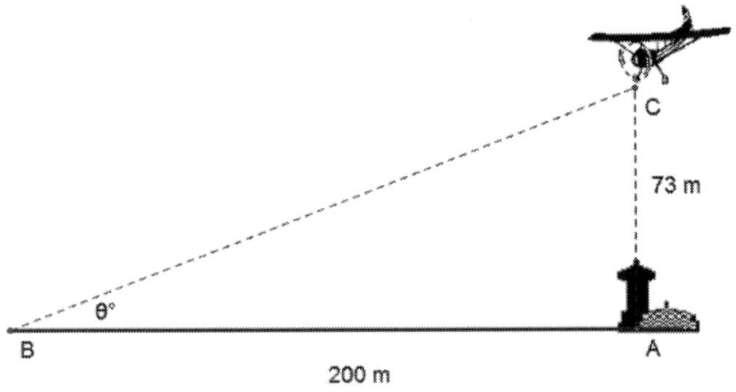

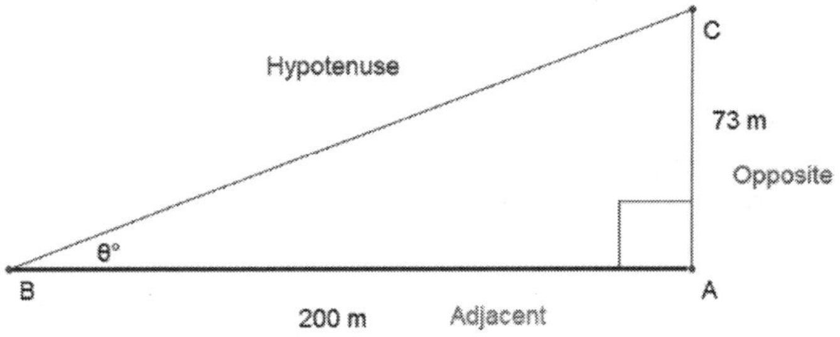

$$\tan\theta^O = \frac{Opposite}{Adjacent}$$

$$\Rightarrow \tan\theta^O = \frac{73}{200}$$

Make sure that your calculator is in Degrees mode

$$\Rightarrow \theta^O = \tan^{-1}\left(\frac{73}{200}\right)$$

This is shift tan or

2nd function tan on your calculator

$$\Rightarrow \theta = 20.052^O$$

$$\Rightarrow \theta = 20.1^O \quad (1\text{ dp})$$

The angle of the glide path of the aircraft is 20.1° (1dp)

USING TAN TO FIND SIDE

Example 2

An aircraft is 200m from its touchdown point, on a glide path of 30˚.

How high above the ground is the aircraft?

Give your answer correct to one decimal place.

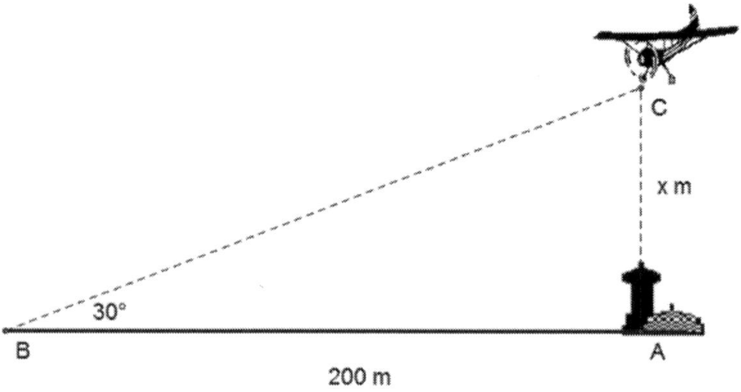

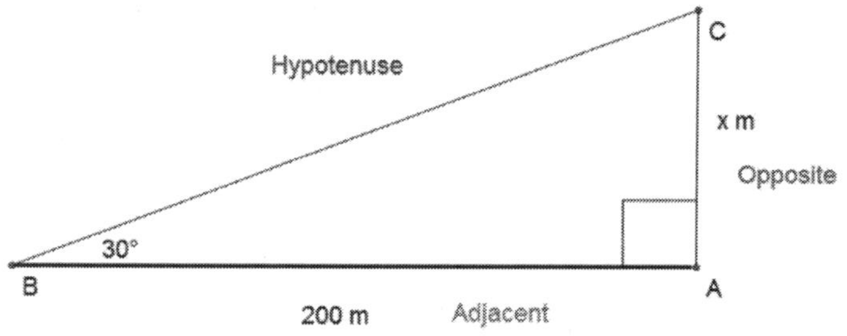

$$\tan 30^{o} = \frac{Opposite}{Adjacent}$$

$$\Rightarrow \tan 30^{o} = \frac{x}{200}$$

Multiply both sides by 200

$$\Rightarrow 200 \times \tan 30^{o} = \frac{x}{\cancel{200}} \times \cancel{200}$$

$$\Rightarrow 200 \tan 30^{o} = x$$

Make sure that your calculator is in Degrees mode

Key in 200 times tan 30 on your calculator

$\Rightarrow x = 115.4700 \quad m$

$\Rightarrow x = 115.5 \;\; m \;\; (\text{ 1 dp})$

The height of the aircraft is 115.5m (1dp)

USING SIN TO FIND SIDE

Example 3

A skier is racing down a 150m long ramp which has an angle of inclination of 30°.

How high above the ground is the starting flag?

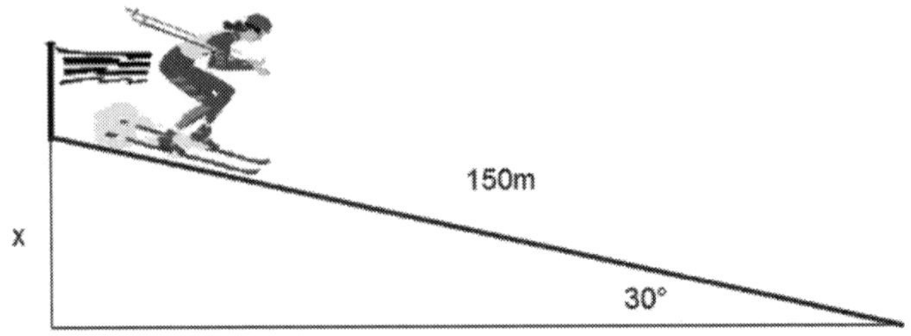

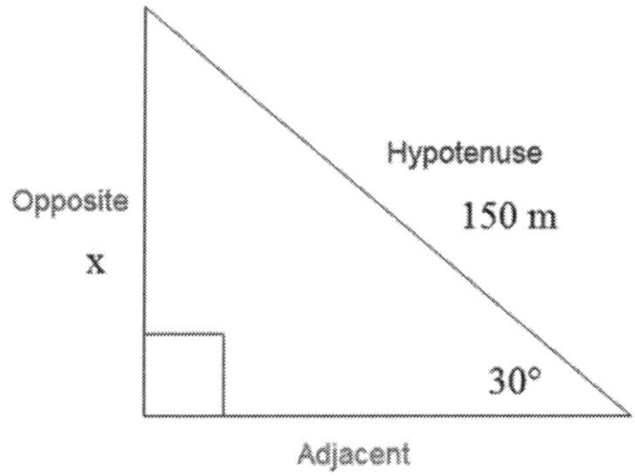

$$\sin 30^o = \frac{Opposite}{Hypotenuse}$$

$$\Rightarrow \sin 30^o = \frac{x}{150}$$

Multiply both sides of the equation by 150

$\Rightarrow 150 \times \sin 30^o = x$

Make sure that your calculator is in Degrees mode

Key in 150 times sin 30 on your calculator

$x = 75 \ \ m$

The starting flag is 50m above ground.

USING SIN TO FIND ANGLE

Example 4

A skier is racing down a ramp which is 200m long.

The start of the ramp is 100m above ground level.

What is the value of θ°, the angle of inclination of the ramp?

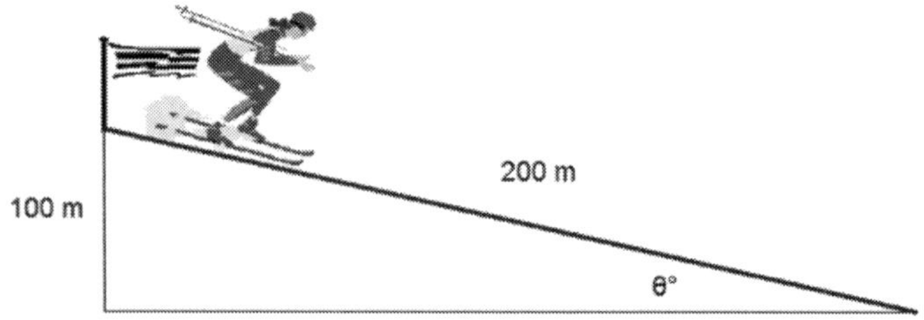

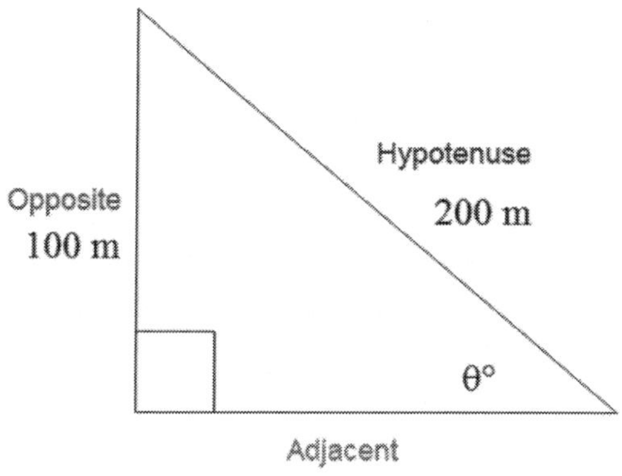

$$\sin\theta^{o} = \frac{Opposite}{Hypotenuse}$$

$$\Rightarrow \sin\theta^{o} = \frac{100}{200}$$

$$\Rightarrow \sin\theta^{o} = 0.5$$

Make sure that your calculator is in Degrees mode

$$\Rightarrow \theta^{o} = \sin^{-1}(0.5)$$

This is shift sin or

2nd function sin on your calculator

$$\Rightarrow \theta = 30^{o}$$

The angle of inclination of the ramp is 30˚.

USING COS TO FIND SIDE

Example 5

A ship is at anchor. The chain is 150 m long and makes an angle of 30˚ from the anchor point to the seabed. The anchor point is 5m above the water level. How deep in the water is the anchor ? Give your answer to the nearest metre.

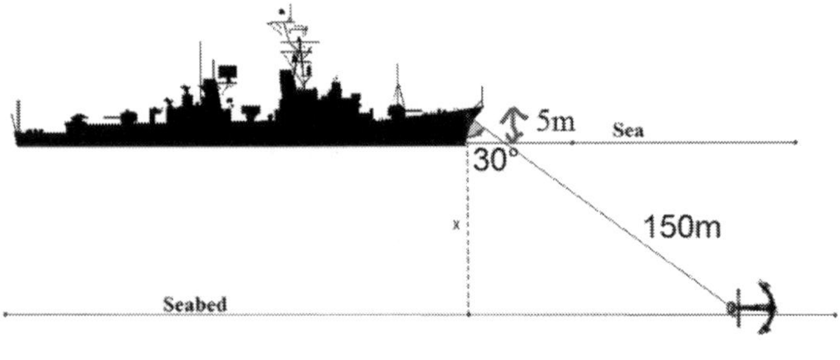

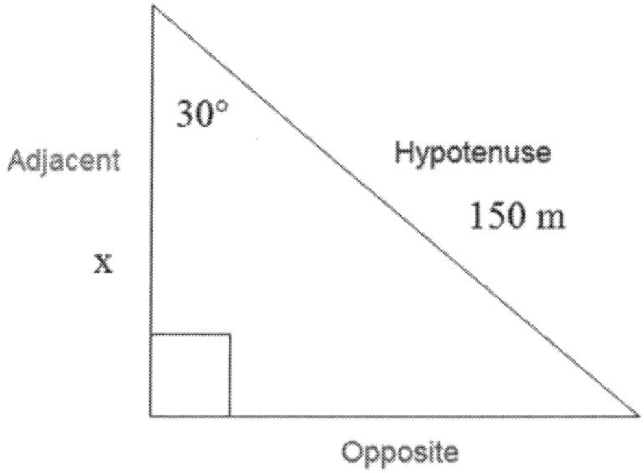

$$\cos 30^{o} = \frac{Adjacent}{Hypotenuse}$$

$$\Rightarrow \cos 30^{o} = \frac{x}{150}$$

$$\Rightarrow 150 \times \cos 30^{o} = x$$

$x = 129.9038 \quad m$

$x = 130\ m$ to the nearest metre

but anchor is already 5m above water line

Depth of anchor $= 130m$ -5m

=125m(nearest metre)

USING COS TO FIND ANGLE

Example 6

A ship is taut at anchor. The chain is 150 m long and lies 75m from the anchor point. What is the value of θ°, the angle from the seabed to the anchor point at the ship's bow ?

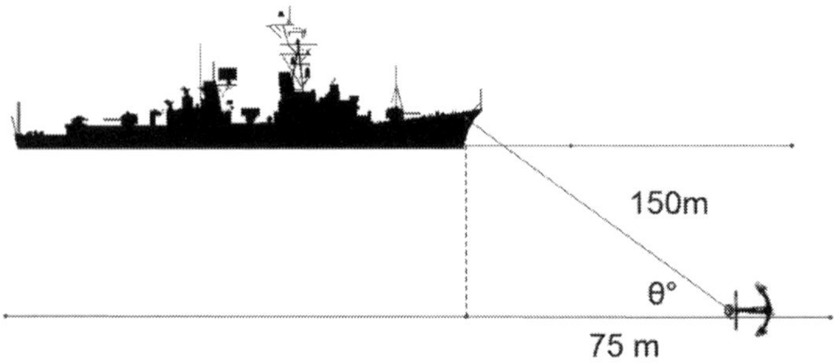

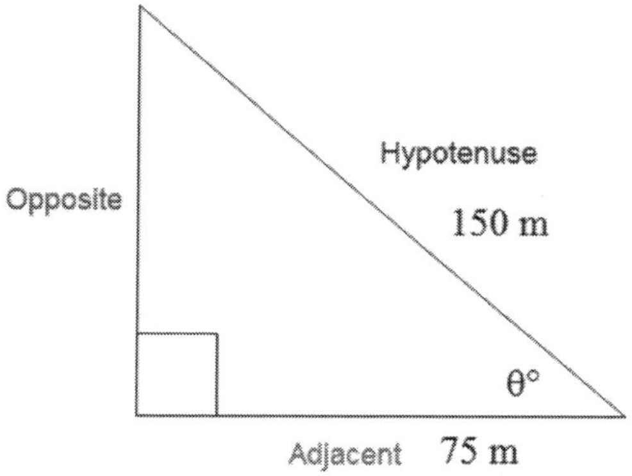

$$\cos\theta^{O} = \frac{Adjacent}{Hypotenuse}$$

$$\Rightarrow \cos\theta^{O} = \frac{75}{150}$$

$$\Rightarrow \cos\theta^{O} = \frac{1}{2}$$

Make sure that your calculator is in Degrees mode

$\Rightarrow \theta^{O} = \cos^{-1}(0.5)$

This is shift cos or 2nd function cos on your calculator

$\Rightarrow \theta = 60^{O}$

The angle from the seabed to the ship's bow is 60°

WHICH RATIO SHOULD BE USED ?

The following recipe can be used for trig problem solving questions :-

1. Find and sketch the triangle

2. Mark the right angle

3. Identify and mark angle to be used / found

4. Label Opposite, Hypotenuse and Adjacent

5. Write out

$$S\frac{O}{H} \qquad C\frac{A}{H} \qquad T\frac{O}{A}$$

6. Two Tick Test:-

   Tick what you know

   Tick what you want

Look for two ticks and use that ratio

7. Write out ratio and solve.

8. Write down solution

Example 7

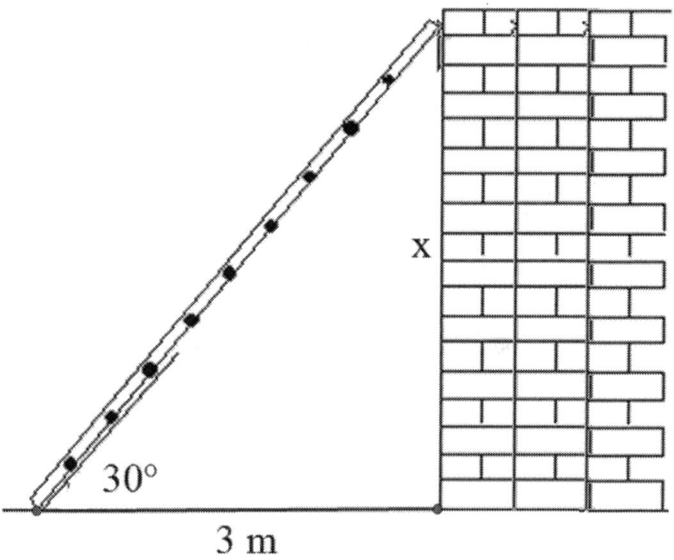

How high up the wall is the ladder ?

(Give your answer to 2d.p.)

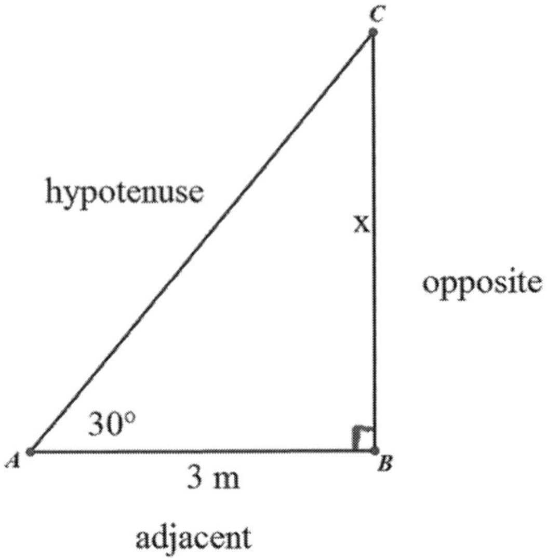

Tick what you know Adjacent

Tick what you want Opposite

$S\frac{O✓}{H}$ $C\frac{A✓}{H}$ $T\frac{O✓}{A✓}$

Look for two ticks and use that ratio.

(Here, use tan)

$$\tan 30^{o} = \frac{Opposite}{Adjacent}$$

$$\Rightarrow \tan 30^{o} = \frac{Opposite}{3}$$

$$\Rightarrow 3\tan 30^{o} = Opposite$$
$$\Rightarrow Opposite = 1.73$$

The ladder is 1.73m (2dp) up the wall.

FINDING THE AREA OF A TRIANGLE

What is the area of this triangle?

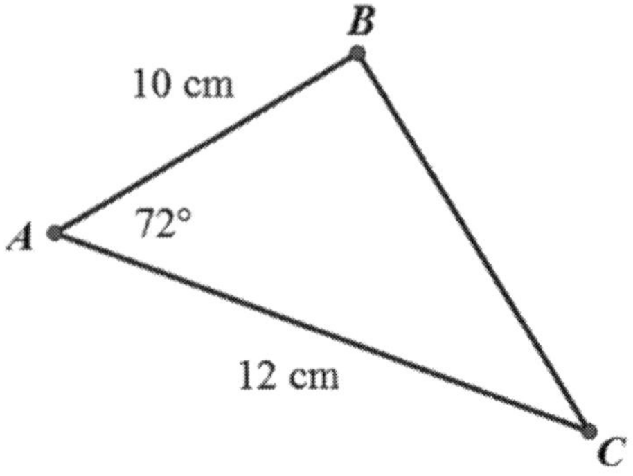

Method 1

Draw perpendicular BD.

Use basic trigonometry to find length of BD.

(This is the height of the triangle.)

Then use Area = 1/2 x base x perpendicular height

So,

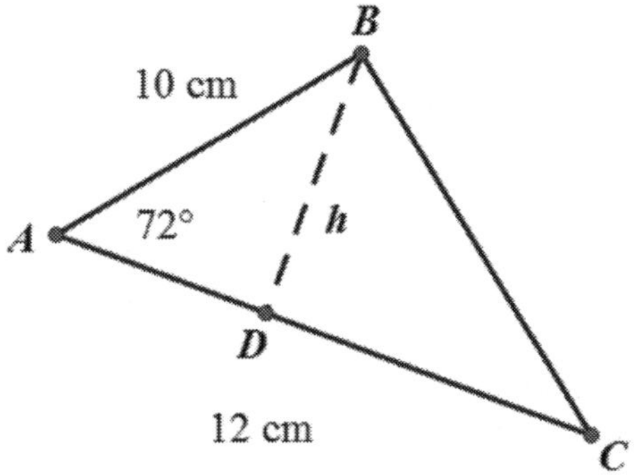

$$\sin 72° = \frac{h}{10}$$

$$\Rightarrow h = 10\sin 72°$$

$$\Rightarrow h = 9.511\,cm$$

A = 1/2 x base x height

A = 1/2 x 12 x 9.511

A = 57.063391

A = 57.1 cm^2 (1 dp)

Method 2

Name sides opposite vertices with their lower case letters.

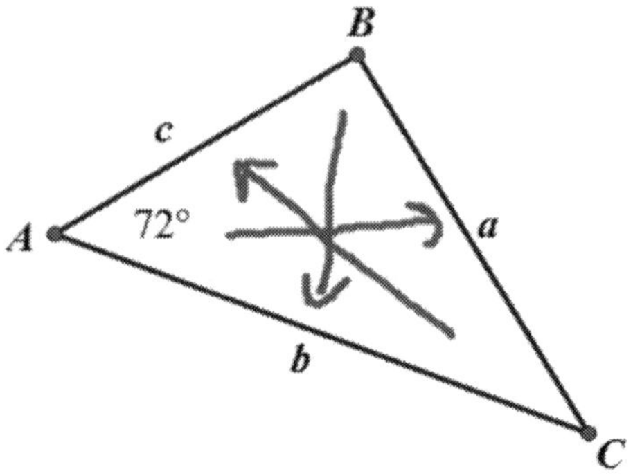

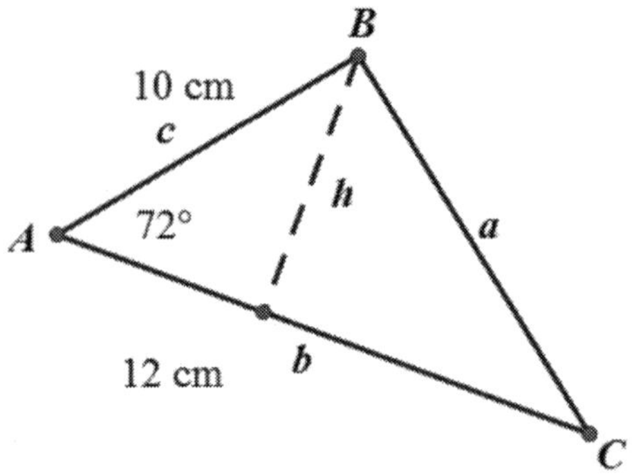

$$\sin 72^\circ = \frac{h}{10}$$

$$\Rightarrow h = 10\sin 72^\circ$$

$$\Rightarrow h = 9.511$$

but sin A $= \dfrac{\text{h}}{10} = \dfrac{\text{h}}{\text{c}}$

$$\Rightarrow h = c\text{sin A}$$

$$Area = 1/2 \times base \times perpendicular\ height$$

$$= 1/2 \times b \times c\sin A$$

$$= 1/2 \times 12 \times 10\sin 72^\circ$$

$$= 57.0633\ cm^2$$

$$= 57.1\ cm^2\ (1dp)$$

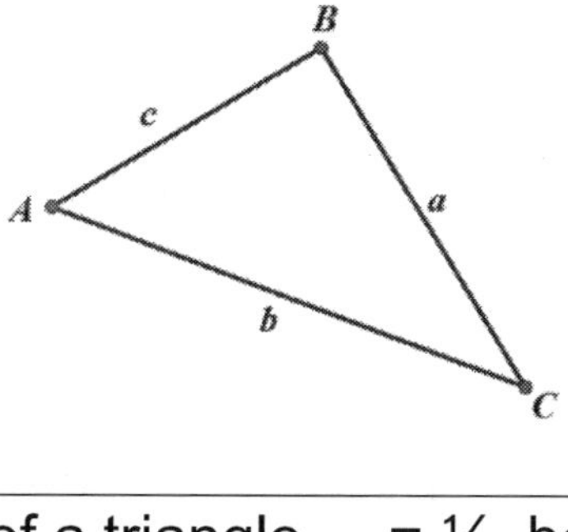

Area of a triangle	= ½ bc sinA
	= ½ ac sinB
	= ½ ab sinC

Example 8

A Hollywood film set is being constructed in the desert. The painter is approaching the square based pyramid from behind a sand dune.

The base is 11m, the slant height is 9m and the angle between the base and slant height is 52°.

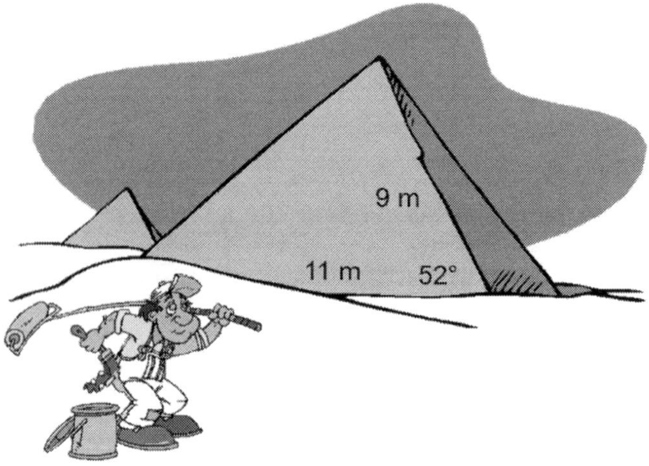

Calculate the area of the face of the pyramid shown.

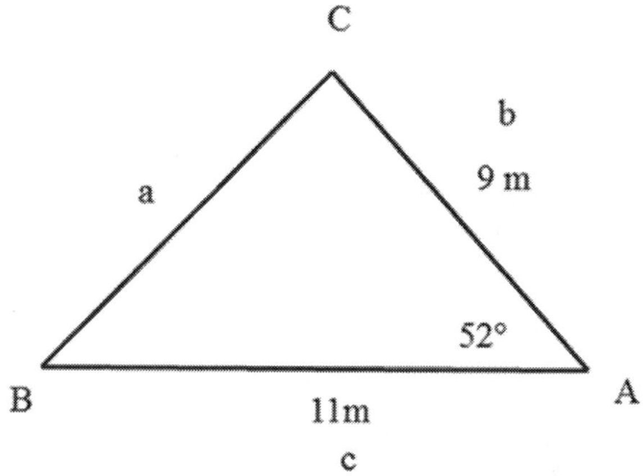

$$\text{Area}_{\text{V}} = \frac{1}{2} bc \sin A$$

$$\text{Area}_{\text{V}} = \frac{1}{2} \times 9 \times 11 \times \sin 52°$$

$$\text{Area}_{\text{V}} = \frac{1}{2} \times 78.013606$$

$$\text{Area}_{\text{V}} = 39.0065$$

The area of the face shown is $39m^2$

THE SINE RULE

What is the length of AC ?

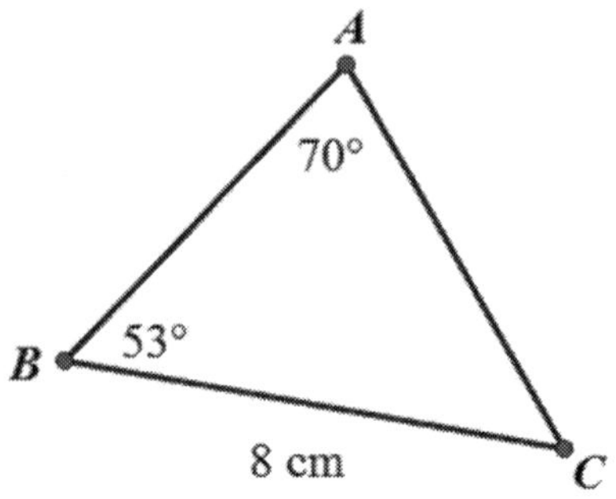

1. Name sides opposite vertices with their lower case letters.
2. Draw perpendicular CD.
3. Use basic trigonometry to find length of CD

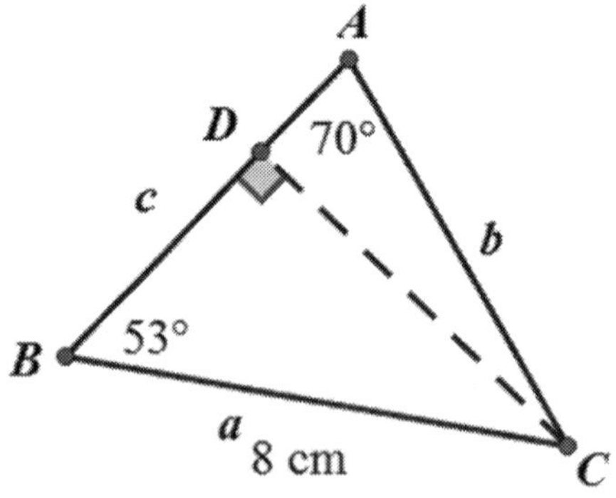

$$\sin 53° = \frac{CD}{8} \quad \text{and} \quad \sin 70° = \frac{CD}{b}$$

$$\Rightarrow CD = 8\sin 53° \qquad \Rightarrow CD = b\sin 70°$$

So

$$8\sin 53° = b\sin 70°$$
$$\frac{8\sin 53°}{\sin 70°} = b$$
$$b = 6.799cm$$

Using the letters

$$CD = a\sin B \quad \text{and} \quad CD = b\sin A$$

$$a\sin B = b\sin A$$
$$\frac{a\sin B}{\sin A} = b$$
$$\frac{a}{\sin A} = \frac{b}{\sin B}$$

Which gives The Sine Rule

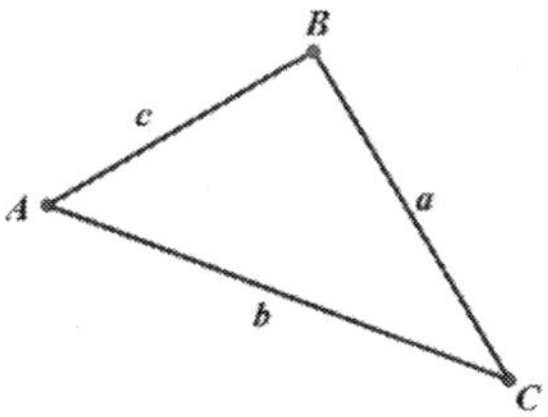

$$\boxed{\frac{a}{\sin A} = \frac{b}{\sin B} = \frac{c}{\sin C}}$$

Example 9

A helicopter is approaching an oilrig. The distance from the helicopter to the oilrig is 15 km.

From the bridge of the tanker, the oil rig bears due East and the helicopter bears 040° at a distance of 10 km.

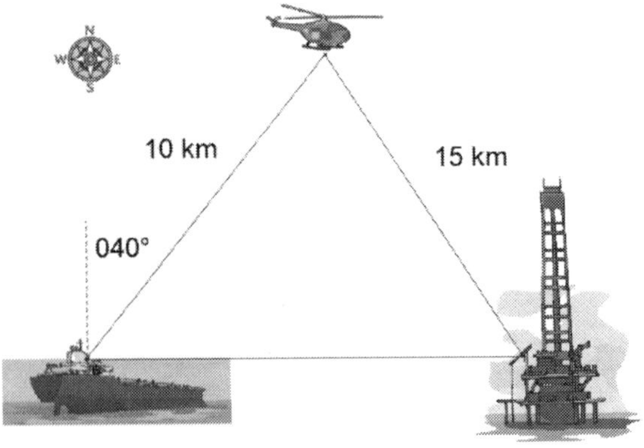

What is the bearing of the helicopter from the oil rig ?

Give your answer correct to the nearest degree.

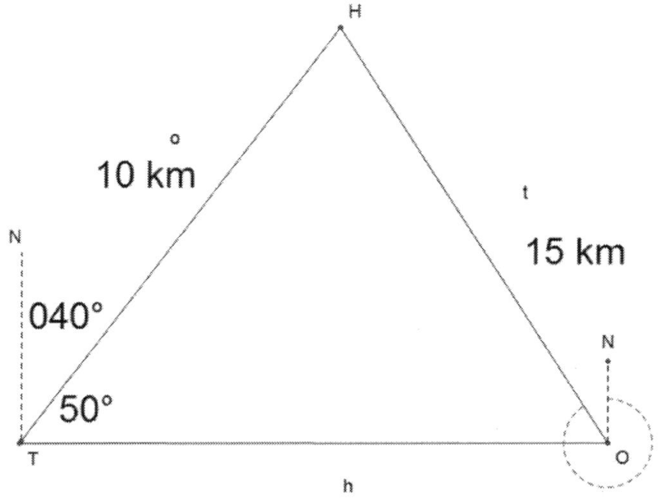

The bearing of H from O = $H\hat{O}T + 270°$

To find $H\hat{O}T$, check to see if Sine Rule works:-

$$\frac{a}{\sin a} = \frac{b}{\sin b} = \frac{c}{\sin c}$$

$$\Rightarrow \quad \frac{h}{\sin H} = \frac{o}{\sin O} = \frac{t}{\sin T}$$

$$\Rightarrow \quad \frac{h}{\sin H} = \frac{10}{\sin O} = \frac{15}{\sin 50}$$

Sine Rule works

$$\frac{10}{\sin O} = \frac{15}{\sin 50}$$

$$10 \times \sin 50 = 15 \times \sin O$$

$$\Rightarrow \quad \frac{10 \times \sin 50}{15} = \sin O$$

$$\Rightarrow \quad O = \sin^{-1}\left(\frac{10 \times \sin 50}{15}\right)$$

$$\Rightarrow \quad O = 30.710°$$

Bearing $= \ 30.710° + 270° = 300.710°$

The helicopter bears 301°(nearest degree) from the oilrig.

THE COSINE RULE

What is the length of BC ?

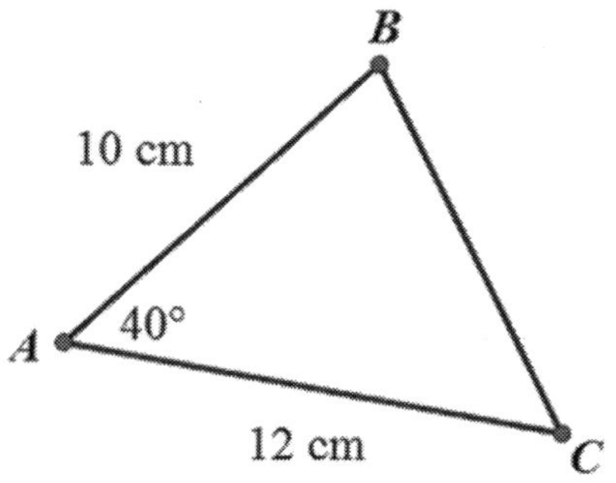

1. Name sides opposite vertices with their lower case letters.

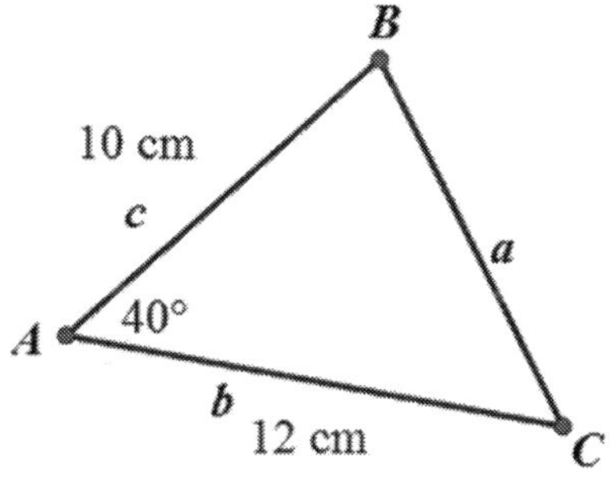

2. Try the Sine Rule

$$\boxed{\frac{a}{\sin A} \overset{}{\underset{\checkmark}{=}} \frac{b \checkmark}{\sin B} = \frac{c \checkmark}{\sin C}}$$

No pairs – so can't use Sine Rule

3. Draw perpendicular BD.
4. Use basic trigonometry to find lengths of BD and AD

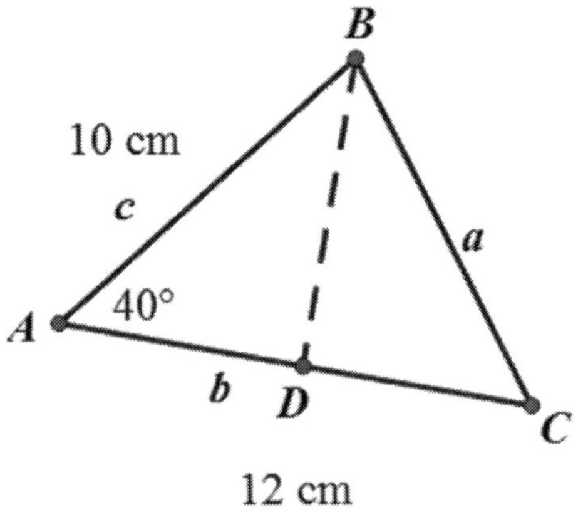

$$\sin 40° = \frac{BD}{10}$$
$$\Rightarrow BD = 10 \sin 40°$$
$$\Rightarrow BD = c \sin A$$

$$\cos 40° = \frac{AD}{10}$$
$$\Rightarrow AD = 10 \cos 40°$$
$$\Rightarrow AD = c \cos A$$

5. Find length DC

$$DC = 12 - AD$$

$$\Rightarrow DC = 12 - c\cos A$$

$$\Rightarrow DC = b - c\cos A$$

6. Use Pythagoras' Theorem to find BC

$$(BC)^2 = (BD)^2 + (DC)^2$$

$$\Rightarrow a^2 = (c\sin A)^2 + (b - c\cos A)^2$$

$$\Rightarrow a^2 = c^2\sin^2 A + (b - c\cos A)(b - c\cos A)$$

$$\Rightarrow a^2 = c^2\sin^2 A + b^2 - bc\cos A - bc\cos A + c^2\cos^2 A$$

$$\Rightarrow a^2 = c^2\sin^2 A + b^2 - 2bc\cos A + c^2\cos^2 A$$

$$\Rightarrow a^2 = c^2\left(\sin^2 A + \cos^2 A\right) + b^2 - 2bc\cos A$$

$$\Rightarrow a^2 = c^2 + b^2 - 2bc\cos A$$

$$a^2 = 12^2 + 10^2 - 2 \times 12 \times 10 \times \cos 40°$$

$$a^2 = 144 + 100 - 240\cos 40°$$

$$a^2 = 244 - 240\cos 40°$$

$$a^2 = 60.14933365$$

$$a = \sqrt{60.14933365}$$

$$a = 7.755$$

$$a = 7.8\ cm \quad \text{(nearest cm)}$$

This is known as the Cosine Rule

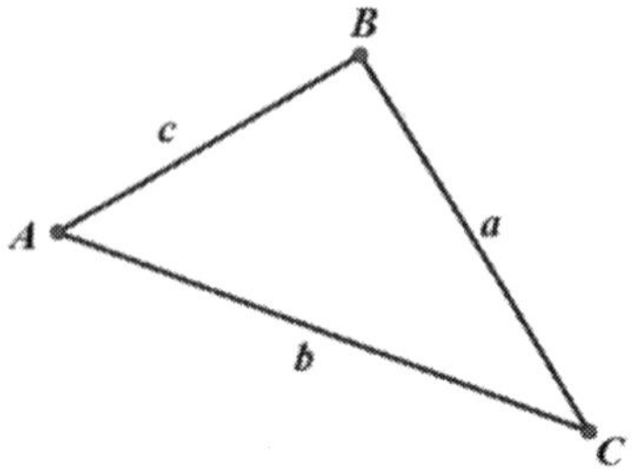

$$\boxed{a^2 = c^2 + b^2 - 2bc\cos A}$$

Re-arranging gives

$$a^2 = c^2 + b^2 - 2bc\cos A$$

$$2bc\cos A + a^2 = c^2 + b^2$$

$$2bc\cos A = c^2 + b^2 - a^2$$

$$\cos A = \frac{c^2 + b^2 - a^2}{2bc}$$

$$\boxed{\cos A = \frac{c^2 + b^2 - a^2}{2bc}}$$

Example 10

A satellite is being tracked by an observatory telescope and a transmitter.

When the angle of elevation of the telescope is 40°, the satellite is known to be a distance of 6500 km from the observatory.

The distance between the observatory and the transmitter is 50 km, at a level height.

What is the distance of the satellite from the transmitter?

Give your answer to the nearest kilometer.

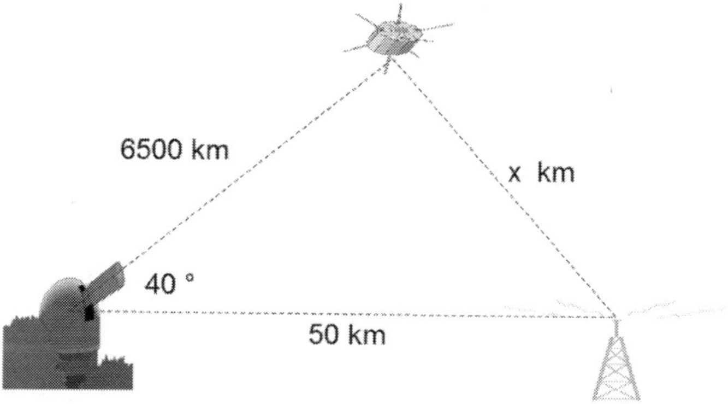

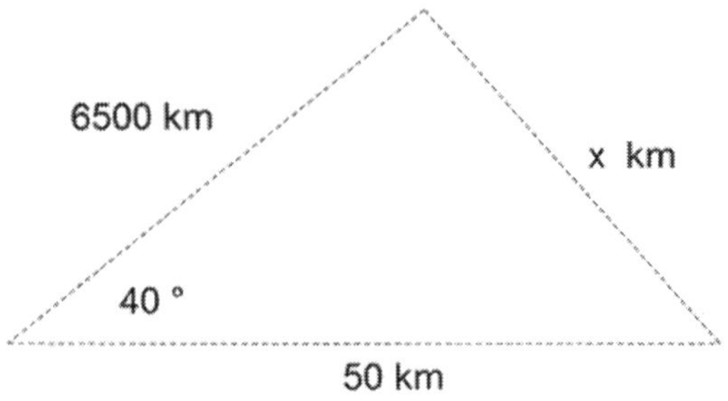

Use Cosine rule

$$\boxed{a^2 = c^2 + b^2 - 2bc\cos A}$$

$$a^2 = 6500^2 + 50^2 - 2 \times 6500 \times 50 \times \cos 40°$$

$$a^2 = 42250000 + 2500 - 650000\cos 40°$$

$$a^2 = 42252500 - 650000\cos 40°$$

$$a^2 = 41754571.11$$

$$a = \sqrt{41754571.11}$$

$$a = 6461.7777$$

$$a = 6462\ km \quad \text{(nearest km)}$$

Example 11

The space shuttle is tied down by two 120 m long chains.

The distance between the tie down points on the ground is 100 m.

What is the value of angle θ° ? Give your answer correct to 1 decimal place.

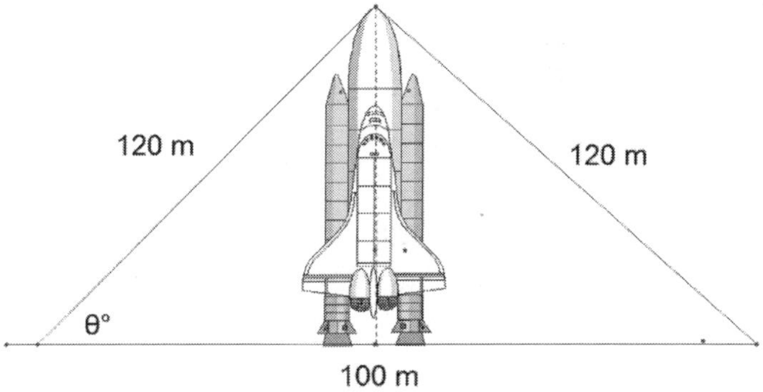

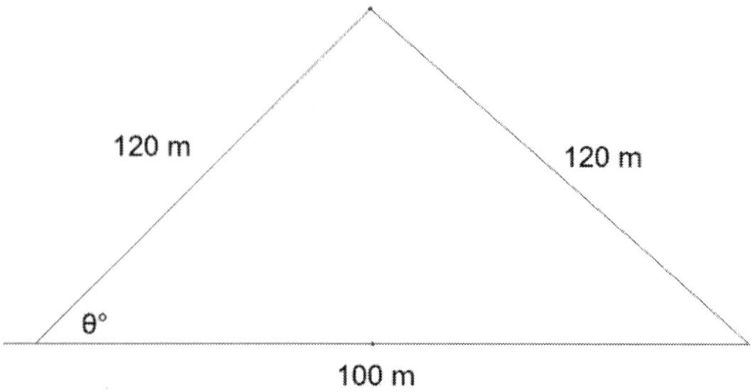

$$\boxed{\cos A = \frac{c^2 + b^2 - a^2}{2bc}}$$

$$\cos\theta = \frac{120^2 + 100^2 - 120^2}{2 \times 100 \times 120}$$

$$\cos\theta = \frac{14400 + 10000 - 14400}{2 \times 100 \times 120}$$

$$\cos\theta = \frac{10000}{24000}$$

$$\cos\theta = \frac{10}{24}$$

$$\theta = \cos^{-1}\left(\frac{10}{24}\right)$$

$$\theta = 65.37568$$

$$\theta = 65.4^\circ \quad (1dp)$$

WHICH RULE TO USE

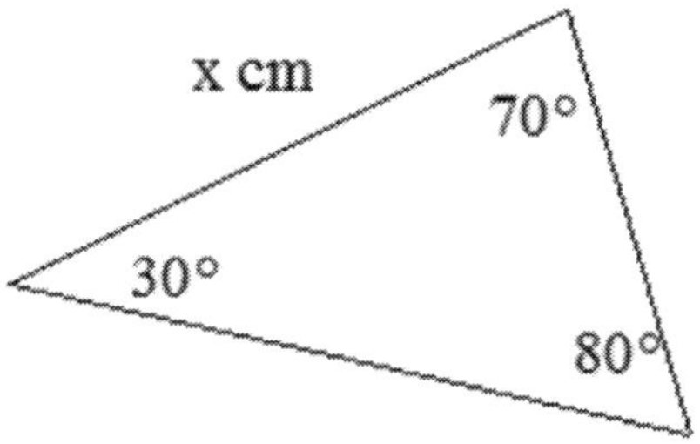

AAA – 3 angles

Not enough information

AAS – 2 angles, 1 side

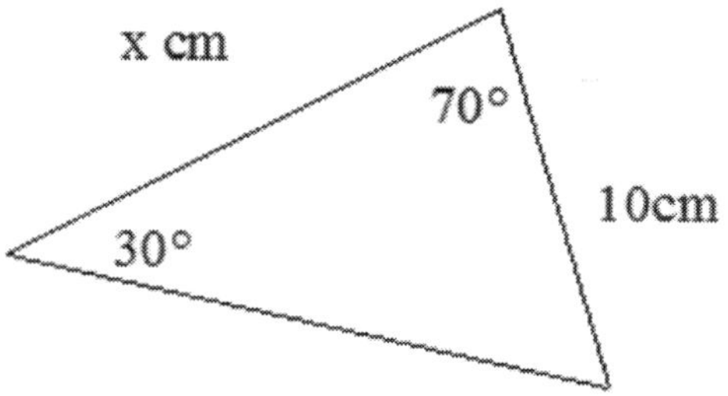

Use Sine Rule

ASS – 1 angle (not included) , 2 sides

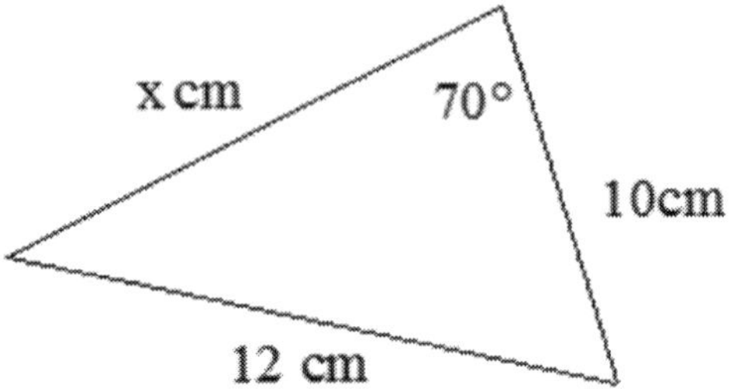

Ambiguous case: Either Use Sine Rule twice, or use Cosine Rule with quadratic formula.

SAS – Included angle, 2 sides

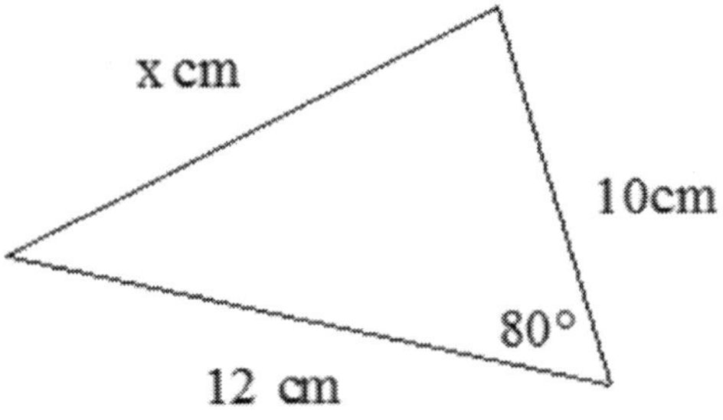

Use Cosine Rule

SSS – 3 sides

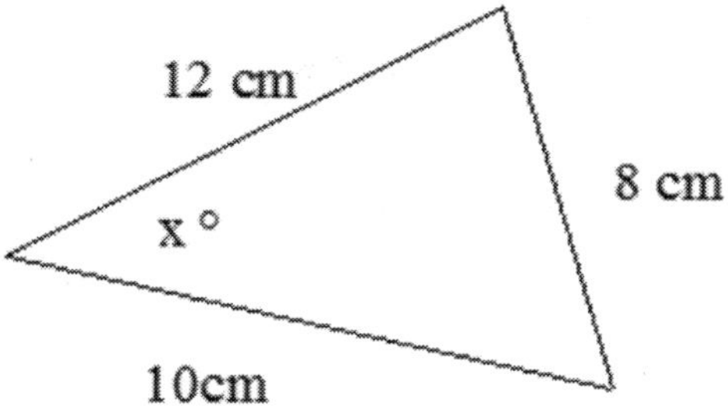

Use Cosine Rule

BOOKS BY THE AUTHOR

The **Pooch Puzzles** series is a collection of printed crosswords, sudokus, word searches and other puzzles - some educational, some just for fun.

https://www.amazon.co.uk/dp/B0BNP17L8N

The **Revision** series is a collection of notes and examples to help with revision.

https://www.amazon.co.uk/gp/product/B08RDYXHLL/

The **Drill Questions** series provides material to help improve fluency in skills.

https://www.amazon.co.uk/dp/B08RF2H3ND

See all titles at

https://www.amazon.co.uk/Alexander-Forrest/e/B07P8J6PHF/

or visit Maths Mutt
http://www.mathsmutt.co.uk

http://www.scruffs.shetland.co.uk/store.html

Made in the USA
Monee, IL
04 November 2026

67375800R00026